LITTLE CRAFT
BOOK SERIES

Tissue Paper Creations

CHESTER JAY ALKEMA

Associate Professor of Art
Grand Valley State College
Allendale, Michigan

Photographs by the Author

STERLING PUBLISHING CO., INC. NEW YORK
Oak Tree Press. Co., Ltd. London & Sydney
SAUNDERS OF TORONTO, Ltd., Don Mills, Canada

Little Craft Book Series

Balsa Wood Modelling
Bargello Stitchery
Beads Plus Macramé
Big-Knot Macramé
Candle-Making
Cellophane Creations
Ceramics by Slab
Coloring Papers
Corrugated Carton Crafting
Crafting with Nature's Materials
Creating Silver Jewelry with Beads
Creating with Beads
Creating with Burlap
Creating with Flexible Foam
Creative Lace-Making with Thread and Yarn
Cross Stitchery
Curling, Coiling and Quilling
Enamel without Heat

Felt Crafting
Finger Weaving: Indian Braiding
Flower Pressing
Folding Table Napkins
Hooked and Knotted Rugs
How to Add Designer Touches to Your Wardrobe
Ideas for Collage
Junk Sculpture
Lacquer and Crackle
Leathercrafting
Macramé
Make Your Own Elegant Jewelry
Making Paper Flowers
Making Shell Flowers
Masks
Metal and Wire Sculpture
Model Boat Building

Monster Masks
Nail Sculpture
Needlepoint Simplified
Off-Loom Weaving
Organic Jewelry You Can Make
Potato Printing
Puppet-Making
Repoussage
Scissorscraft
Scrimshaw
Sculpturing with Wax
Sewing without a Pattern
Starting with Stained Glass
String Things You Can Create
Tole Painting
Trapunto: Decorative Quilting
Whittling and Wood Carving

The author and publishers wish to extend their appreciation and indebtedness to the many people who contributed in various ways to this book: the children of Wyoming (Michigan) Parkview School whose tissue-paper creations are pictured here; the Art Education students of Grand Valley State College for the many ideas and examples they provided, and for their assistance in photographing the art in the book; and the editors and publishers of *School Arts* and *Arts and Activities* for the use of the author's material they originally published.

Copyright © 1973 by Sterling Publishing Co., Inc.
419 Park Avenue South, New York, N.Y. 10016
Distributed in Canada by Saunders of Toronto, Ltd., Don Mills, Ontario
British edition published by Oak Tree Press Co., Ltd., Nassau, Bahamas
Distributed in Australia and New Zealand by Oak Tree Press Co., Ltd.,
P.O. Box 34, Brickfield Hill, Sydney 2000, N.S.W.
Distributed in the United Kingdom and elsewhere in the British Commonwealth
by Ward Lock Ltd., 116 Baker Street, London W 1
Manufactured in the United States of America
All rights reserved
Library of Congress Catalog Card No.: 73-83454
ISBN 0-8069-5288-1   UK 7061-2468-5
5289-X

Contents

Before You Begin 4
    Make a Tissue-Cord Drawing

Crushed Tissue Figures 6

Baubles 7

Tissue-Parchment Greeting Cards 8

Tissue Dye Transfers 9

Laminated Tissue Creations 14
    Collages . . . Containers . . . Easter Eggs

Stencilled Designs 22

Tissue-Tagboard Sculpture 27

Tissue-Carton Sculpture 31

Flowers and "Mops" 34

Tissue-Chicken Wire Sculpture 35

Tissue-Wire Sculpture 37

Transparencies 39
    Mobiles . . . "Stained-Glass" Windows . . . Insects

Bleached Tissue Creations 43
    Bleach Paintings . . . Bleach Printing

Printing on Tissue 46

Index 48

Before You Begin

Tissue paper has certainly come a long way from a "gauzelike paper, used to protect engravings in books, to wrap up delicate articles, etc.," which is how Webster defines it!

Today, tissue paper is one of the most exciting craft media to explore. Available in a multitude of brilliant, intense colors, tissue will motivate you to creativity the moment you open up a package. And that is the reason for this book—to open up the world of possibilities awaiting you in that package.

For example, some of the things you can *do* with it are:

- lay layer on layer, color on color
- sponge on water and transfer the color to other papers
- bleach-paint on it
- draw on it with soft-tip ink markers, crayons, etc.
- crush it into shapes
- accordion-pleat it into fantasies
- potato-print on it
- sculpture with it
- combine it with scrap materials.

Some of the things you can *make* with it are:

- abstract designs
- representational designs
- collages
- containers
- "Easter eggs"
- stencilled designs
- party hats
- mobiles
- stabiles
- animals
- flowers
- Christmas tree decorations
- "stained-glass" windows
- gift wrappings
- murals
- greeting cards.

All of the methods and the many ideas offered you in these pages are designed to lead you into your own avenues of exploration. Once you get started, you'll come up with so many ideas you may have to write a book of your own!

Make a Tissue-Cord Drawing

If you have never "worked" with tissue, take your first plunge by making a tissue-cord "drawing."

You might want to make an abstract design to begin or a representational one such as the simple flower arrangement in Illus. 1. On a piece of paper make a rough outline of your chosen subject as a general guide to follow.

Then, cut long strips, about 1 inch wide, from a variety of different-colored tissue papers. For example, in Illus. 1, sunny yellow and bright red form the petals, and soft green the stems and leaves.

Now select a cord-like material—string, yarn or even clothesline. Wind the tissue strips around the cord and form them into the various elements of your design. As you finish each part, glue it onto a colorful background of construction paper, foil, tagboard, etc.

There, wasn't that easy? All of the projects that follow are just as easy to do, and the results even *more* stunning.

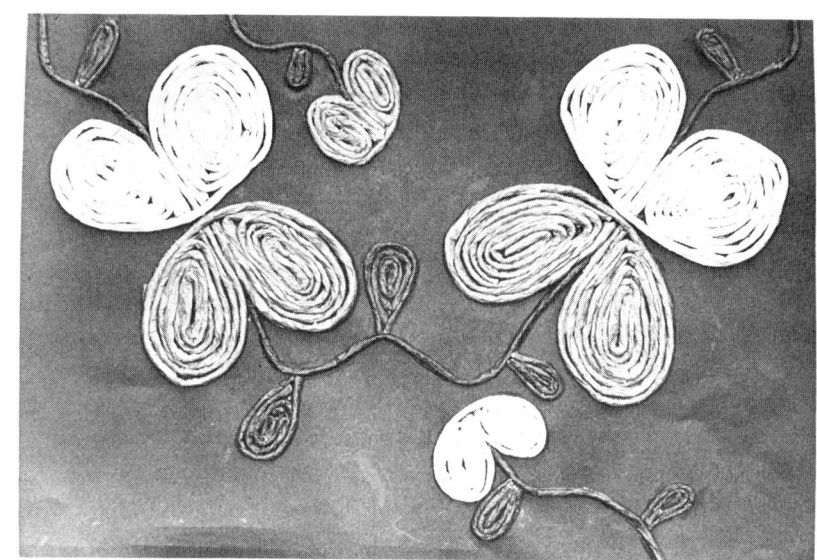

Illus. 1. This tissue-cord flower "drawing," composed of bright colors, would make an ideal wall hanging for your kitchen.

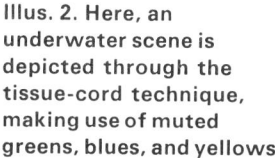

Illus. 2. Here, an underwater scene is depicted through the tissue-cord technique, making use of muted greens, blues, and yellows.

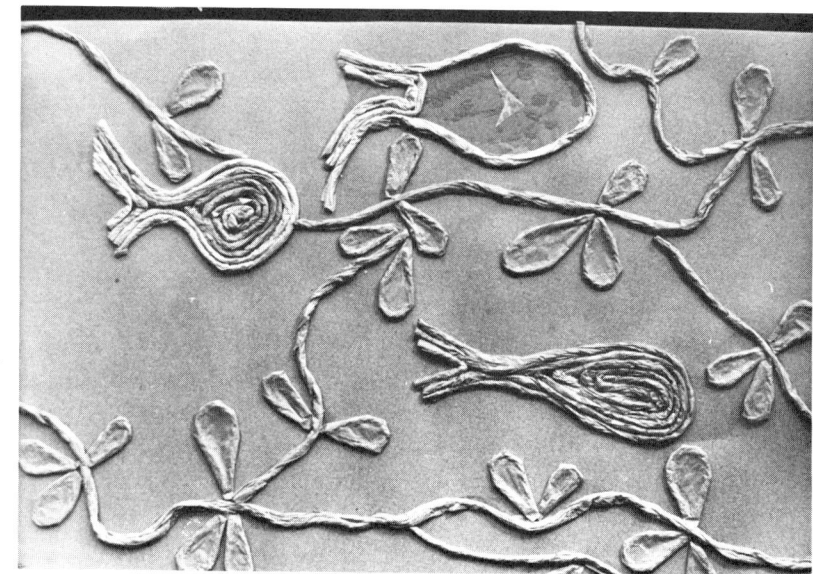

Illus. 3. These charming figures are made of tissue inside tissue! Balls and bunches of tissue are formed into shape with string and then given a "garment" of more tissue.

Crushed Tissue Figures

The happy-looking trio in Illus. 3 are chock-full of tissue! Crush plain white tissue into large "balls." Then use string to tie off the balls into whatever shapes you want.

You can now poke pipe cleaners, toothpicks or wire into the crushed balls to serve as legs, arms, or whatever parts your figure requires. When you are satisfied with the shape of the creature, cover the crushed white tissue with dabs of glue, and place pieces of colored tissue around the parts to completely cover the crushed paper and string.

The outer-space man on the left has pipe-cleaner legs and a tissue-covered wire tail. Tempera paints were used to create his eyes and mouth. The tired-looking mouse in the middle has cylindrical arms, legs and tail made of rolled tissue that was glued onto the crushed-tissue body. The tiny elephant on the right was made from one long crushed ball. String is used to section off his legs, trunk, and ears.

Baubles

Baubles can be anything you want them to be. The bauble in Illus. 4 was used as a Christmas decoration. The shape was cut out of a piece of black construction paper (not folded) and the center removed and lined with colorful tissue. The two side pieces were made the same way. Christmas glitter adorns the black areas.

The party decoration in Illus. 5 has a tagboard

Illus. 5. An elegant bauble such as this makes a perfect party decoration.

base. Colored tissue paper is held down by aluminum foil strips pasted along the edges. Sections cut from a plastic doily are glued over the tissue paper, and glass beads adorn the finished design.

Another kind of bauble is shown in color Illus. 31. Using photographs in this way, you can turn your baubles into keepsakes for friends or relatives.

Illus. 4. Baubles can be any shape and size you wish, depending upon what you are going to do with them.

Tissue-Parchment Greeting Cards

To create a parchment greeting card, cut a square or rectangular shape from waxed paper. Submerge such objects as leaves, pressed flowers, colored yarn, thread, string, sequins, and so on, in a bath of 1 part white glue and 1 part water. Arrange the objects in any way you please on the waxed paper.

Cover the waxed paper and applied objects with a sheet of white or light-colored tissue and saturate the tissue with white glue which you apply with a brush. After the card has thoroughly dried, press it, using a cool iron, between two sheets of paper.

Illus. 6. Tissue-parchment greeting cards are as much fun to make as they are to receive! Simply bathe whatever objects you wish in glue and water, place on waxed paper, and cover with tissue. Here pressed leaves form the left-hand card and evergreen twigs the right-hand card.

Tissue Dye Transfers

One of the easiest and most enjoyable ways to create with tissue is making a dye-transfer design. You might use an unplanned approach or a planned one.

To begin an unplanned design, place a sheet of clean absorbent newsprint on top of a newspaper-covered table (to help keep the table clean). Then either cut or tear out a number of different shapes from a variety of colored tissue papers. Place these at random over the newsprint.

Now dip a sponge into water and apply it generously to the tissue-paper pieces. (If you wish, you might try sponging the newsprint before applying the tissue-paper shapes.) Allow to dry and then remove the tissue paper carefully from the newsprint. You need not cover the entire sheet of newsprint with tissue unless you want to, so that parts of the design will be uncolored.

Now what kind of design did the bleeding dyes create upon the absorbent newsprint? Can

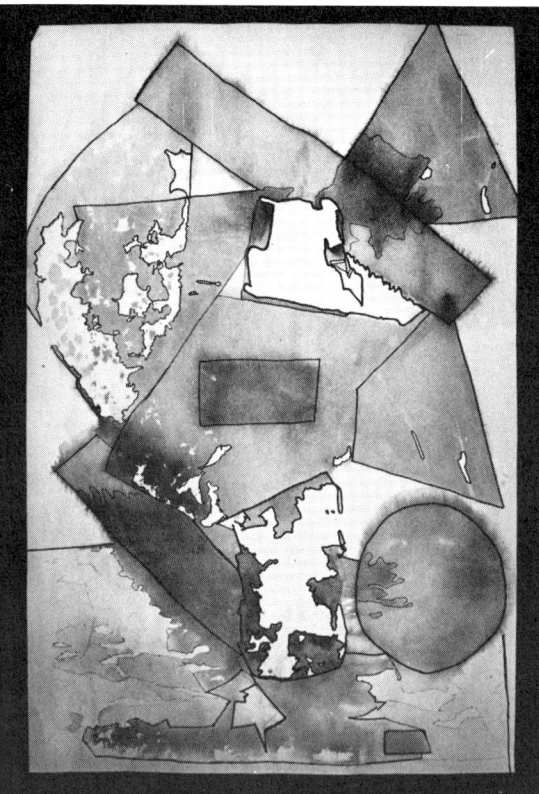

Illus. 7. Free forms are here outlined with pen and ink. Geometric shapes such as the rectangle, triangle, and circle were accidentally created with the dye-transfer method and emphasized with ink.

Illus. 8. First, dip a sponge into water and liberally apply water over tissue-paper shapes. This will cause the tissue-paper colors to bleed onto the underlying newsprint.

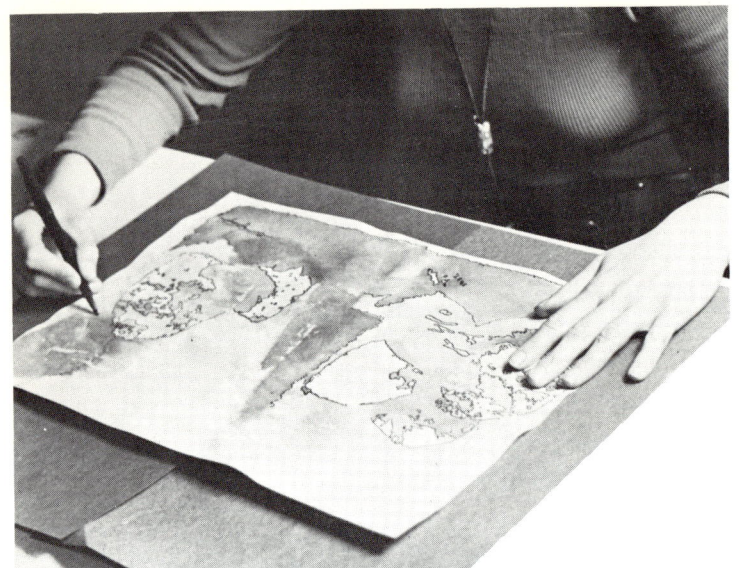

Illus. 9. Here, the artist is outlining certain shapes with pen and ink . . .

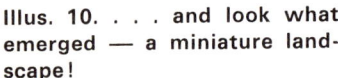

Illus. 10. . . . and look what emerged — a miniature landscape!

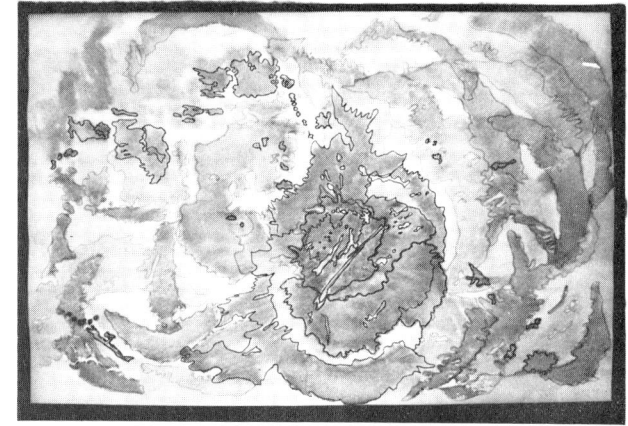

Illus. 11. The resulting forms in this dye-transfer design are purely non-representational, and their dreamy quality is heightened only in certain areas by ink outlines.

you find any forms that suggest a representational object? The artist in Illus. 9 has chosen to use a soft-tip ink marker to outline some of the resulting forms to transform them into recognizable objects. The results of her efforts are in Illus. 10—a garden scene with tall evergreens, carefully pruned shrubs, and a mowed lawn!

In Illus. 11, another free-form dye transfer was exposed to a soft-tip pen outlining; however, here the outlines simply emphasize the existing abstract forms, rather than trying to seek out representational shapes.

In Illus. 12, the resulting areas suggested a cityscape to the artist, so he chose to actually

Illus. 12. A cityscape emerged from a number of orange and red shapes in this design, and the artist created boldly outlined buildings and other structures to complete it.

create a city scene using the basic shapes of the dye-transfer design. As you can see, the dye-transfer technique allows you a great deal of freedom in your tissue-paper creations.

Illus. 16 is an example of how you can plan a dye-transfer design right from the beginning. The artist decided she wanted an underwater scene and carefully cut out a number of fish and seaweed shapes from appropriately colored tissue. She then arranged them on the newsprint, and after the water bath and drying process, *left* the tissue pieces in place and glued them to the newsprint, which resulted in a dye-transfer "collage." The crinkly texture of the dried tissue adds interest to the pen-and-ink outlines which she used to point up the forms.

Illus. 13. Can you see a woodland nymph hiding in this forest of dye-transfer trees?

Illus. 14. Orange, blue, and red dye shapes here suggest a human figure. Notice the added patterns in the coat and foot.

Illus. 15. Imagination can run rampant as accidental shapes emerge in the tissue-dye technique. Here a Halloween ghost jumps over a neatly trimmed hedge. Notice the interesting tree towering over the ghost on the left. This is an example of an unplanned design.

Illus. 16. Here is an example of a planned approach to the tissue-dye transfer method. If you do not want to "leave things to chance," but prefer to know exactly how your design will come out, see page 12.

13

Illus. 17. Many surprising effects are created by laminating tissue paper. Can you see a turtle?

Laminated Tissue Creations

Tissue paper's "see-through" quality opens up another realm of possibilities—by overlaying (or laminating) different colors, unusual effects are achieved. You can utilize the laminating technique for a variety of decorative objects.

Collages

To create a laminated tissue collage, begin by diluting white all-purpose glue with water, in the proportion of 1 part glue to 1 part water. With a brush, paint the diluted glue upon a base of white drawing paper, newsprint, or manila paper.

Cut or tear shapes from tissue and place them on the glued background. Then, paint the diluted glue over this first layer of tissue and

Illus. 18. Bubbles of air caught between the sheets created an unexpectedly interesting texture in this cord-outlined collage.

Illus. 19. If you want to point up certain elements in your laminated collage, use pen and ink.

Illus. 20. Letters, words, and textured shapes cut from magazines effectively combine with tissue-laminated collages.

add more shapes and colors. Continue adding in this way until you feel your design is complete.

When you want a more durable finish for your collages, use commercial latex coatings (such as Art Podge, Decopage, or Mod-Podge) in place of the diluted glue. The surface will be more glossy, more transparent, and longer lasting.

As you can see in Illus. 17, laying layer on layer not only creates new colors, but also creates new *forms*. An interesting textural effect was achieved in Illus. 18 when air bubbles were caught between layers during the lamination process.

As in the dye-transfer technique, you can also use pen and ink or soft-tip ink pens to create new forms or to heighten already existing forms. In Illus. 19 a stem and leaf veins were drawn in with a soft-tip pen to decorate the leaves.

If you want to combine your tissue paper with other materials, magazines provide a rich source of color as well as texture for collages. The collage in Illus. 20 contains magazine patterns, cut into diamond shapes which are echoed in cut-tissue paper forms and pen-and-ink drawings. Notice how, even though this one shape is used throughout, the placement and overlapping of the diamonds creates an exciting variety of parts. (See pages 16, 17, and 20 for a group of laminated collages in full color.)

Containers

So far your laminated creations have been two-dimensional. Now try your hand at creating three-dimensional objects such as containers. These containers will not only be decorative but useful. You can cover a number of different cartons and boxes to produce attractive household necessities such as canisters, letter boxes, knitting holders, and so on.

First, select a nicely shaped container of cardboard, wood, or even glass, such as a wine bottle (see color Illus. 33). Then cover the container

Illus. 21. In decorating his container with the tissue-lamination technique, this artist repeats the shape of the container in his choice of tissue shapes.

Illus. 22. A representational scene is shown in this laminated collage. Sharply pointed waves overlap each other and support two little sailboats. Rugged island shapes fill the background.

Illus. 23. A paper plate was used here to create a circular tissue-laminated design. Shades of yellow and blue tissue form a star shape. Where they overlap, various greens are produced. India ink decorates the star and border of this wall hanging.

Illus. 24. One geometric form, the rectangle, was repeated in many different sizes both in the tissue laminations and in the ink designs. Try using one shape such as this and see if you can create an interesting design, too, simply by varying the sizes and colors as this artist did.

Illus. 25. Here is another example of the use of one shape to create variety. This time it is a tissue-laminated letter "B"!

17

Illus. 26. A variety of tissue-laminated containers is shown here. The tall box has a striking Mediterranean pattern of pink, red, and dark purple overlapping shapes, while the small box on the right has leaf forms created by "bleeding" the tissue and then outlining them with pen and ink.

with white drawing paper, newsprint, or white tissue, using white glue. (Tissue would be best for glass as it is the most flexible.) This white underlayer will intensify the laminated tissue shapes and colors.

Next, you should study the container to determine what tissue shapes will best complement it. For example, in Illus. 21, the artist is using rectangular shapes which echo the rectangular shape of the box. In Illus. 26, a variety of tissue-laminated designs are shown. The flat box in the lower middle is covered with a sheet of solid light blue tissue upon which diamond shapes in magenta and dark purple overlap each other. Some of the diamonds have had their middles cut out as you can see. The box on the left has laminations of black and orange pumpkin shapes overlapping each other. Crayons were used to add decorative lines and pumpkin faces on some of the shapes.

In Illus. 27, the attractive (shoe) box shown here was laminated with snowflake designs, made from folded circles of tissue. First fold the circles in half, then in quarters and then eighths. Cut small pieces from the folds and then open, and you will have a delicate design such as shown. Be sure to vary the cuts in each circle for added interest. Then laminate your snowflakes onto your box.

Illus. 27. Snowflake designs adorn a shoebox

Illus. 28. Two oatmeal boxes have both lost their identities under applications of laminated-tissue shapes.

Illus. 29. Many layers of red and yellow rectangles adorn this striking box. Water was applied generously to the shapes so that their colors would flow down the white paper background.

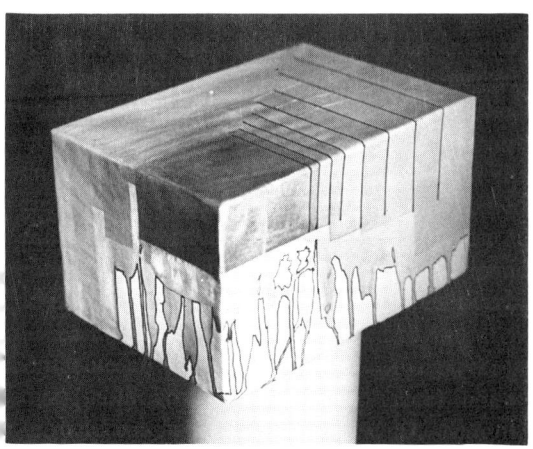

Easter Eggs

Using the tissue-lamination technique, you can create fantastic Easter eggs on balloons! First, paint liquid starch all over a balloon and then apply a layer of tissue-paper strips. Then paint another coat of starch on this first layer and apply more strips. To achieve a sturdy "eggshell," use at least six layers of alternating tissue and starch.

When thoroughly dry, cut one end of the shell away (using a single-edged razor blade), and remove the balloon if you wish. If you cut away a very large hole on the end, you can fill your egg with all kinds of Easter goodies (color Illus. 42).

Then you can decorate your eggs with glitter, rick-rack, yarn, ribbon, pieces of paper doilies, and so on, or add touches here and there with soft-tip pens. See color Illus. 41 for more ideas.

Illus. 30. A tissue-laminated Easter egg in the making! See some finished "eggs" in color on page 24.

Illus. 31. Tissue baubles are fun to make (page 7). A mother used photographs here of her children to make bauble keepsakes for their grandparents. A circular paper doily with the middle removed forms the outer frame for each bauble, and colored tissue covers the middle sections. Christmas glitter provides little star shapes round each photo.

Illus. 32. A combination of letters and photographs intertwine with geometric shapes of tissue paper in this laminated collage.

Illus. 33. Containers of all kinds can be tissue-laminated. In the middle is a nicely shaped wine bottle decorated with orange, green, and yellow circles with Christmas glitter accents.

Illus. 34. This festive container started out as a facial tissue box. Oval green, orange, and yellow shapes overlap the white paper backing. Some colors were allowed to "bleed" when they were applied. Christmas glitter points up the design and adds a lively sparkle to the container.

21

Stencilled Designs

You can now combine your tissue laminations with the crayon-stencil technique to produce unusual and stunning designs. To begin, cut out or tear a number of pieces of tissue in different sizes and colors, but all of the same general shape such as shown in Illus. 36. Laminate them with diluted white glue on a background of either light or dark construction paper, or plain white drawing paper or newsprint. but do not completely cover the background.

Now, from a piece of tagboard, cut out a piece similar in shape to the tissue-paper shapes, as shown in Illus. 35. The cut-out form is called Stencil "A," and the hole left in the tagboard is Stencil "B." (Note that a small form has been cut out of Stencil "A", which is also used in creating this stencil design.)

Illus. 35. Stencil "A" is the piece that you cut out, and Stencil "B" is the hole left in the paper.

Then place Stencil "A" over the tissue-laminated pieces, and rub pastel crayon all round the edges of the stencil. With your finger-tips, rub the pastel crayon onto the surface of the tissue,

Illus. 36. Using a pastel crayon, rub all of edges of the stencil until heavily coated.

Illus. 37. Here is the finished design begun in Illus. 36. Note the new shapes formed by overlapping the stencil forms.

Illus. 38. Blue and magenta arrow shapes, cut from tissue, are laminated on white drawing paper. Black crayon was used to stencil over the tissue to form this unique design.

stroking outwards from the stencil edge. Then re-apply the crayon and move the stencil to another area and repeat. When you are satisfied, do the same with Stencil "B"; however, here you will rub the crayon towards the middle of the hole, which will produce a different result.

Notice in the final design in Illus. 37 that the small shape cut from Stencil "A" has also been

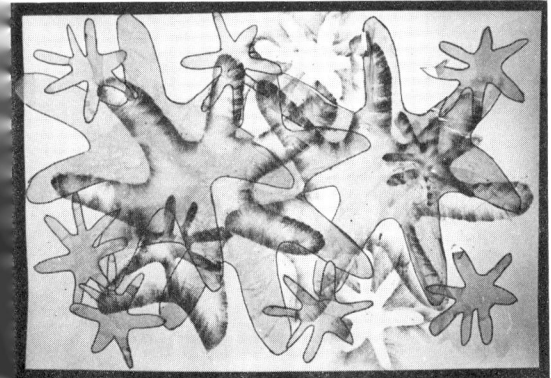

Illus. 39. Free-form stars are laminated and stencilled on yellow construction paper. Green tissue forms are outlined with ink.

Illus. 40. The dye-transfer method, the laminated-tissue technique and the chalk-stencil technique all combined to create this forest.

Illus. 41. These tissue-laminated Easter eggs were formed over balloons! (See page 19.) Soft-tip ink markers and Christmas glitter add decorative touches here and there.

Illus. 42. Rick-rack, sequins and glitter are added to this gigantic egg which you can use to hold Easter goodies.

Illus. 43. Tagboard provides an ideal base for tissue-paper sculpture (page 27). All of the forms here—the locomotive, the giraffe, and the clown—are made from tagboard formed into a variety of geometric shapes, and then covered with colorful tissue paper.

Illus. 45. A shallow tagboard cone was used in constructing this hat. A pinking shears produced jagged edges on the tissue-paper circles and rectangles.

Illus. 44. A large tagboard cone forms the basis for this extravagant tissue-paper-covered party hat. Long tissue-paper streamers flow down from the peak.

25

In Illus. 46, an exciting variety of parts was achieved with the stencil technique. Large blossom shapes were cut from yellow and orange tissue for laminating. Then smaller stencilled blossoms were made to overlap the tissue. India ink was then added to create solid and linear forms and brown crayon fills the background areas. The fiery forms in Illus. 47 are made of red tissue laminated onto black construction paper. The flame-shaped stencils were applied with flesh-colored pastel crayon.

Illus. 46. A delicate flower motif is here produced by the stencil technique.

stencilled onto the background. Be sure to re-crayon your stencils each time you start.

Illus. 46 and Illus. 47 are a study in contrasts!

Illus. 47. Flame-shaped stencils created this exotic design.

Illus. 48.

Illus. 49.

tissue cut into lightning shapes and attached with white glue. Tissue-paper frills adorn the base. The dunce-like hat in Illus. 49 is covered with a variety of tissue-paper shapes—flowers, squares, rectangles, and has long tissue-paper streamers glued to the top of the cone. (See color Illus. 44 for another tall party hat.)

The coolie-type hat in Illus. 50 is made in the same way, but this time, the tagboard was formed into a shallow cone. The pieces of tissue that adorn this hat were trimmed, as you can see, with a pinking shears before being glued onto the tagboard with white glue.

The cone is not the only basic tagboard shape you can fashion into a base for tissue-paper creations. (See color Illus. 43.) For example, in Illus. 55, the lion, the clown, and the spotted pony are all made from cylinders, that is, rectangles of tagboard rolled up and taped together. The mean-looking witch in Illus. 56 has one long

Tissue-Tagboard Sculpture

Tagboard, because of its stiff, firm nature, provides an ideal foundation for tissue-paper overlays.

The fantastic party hats in Illus. 48 and Illus. 49 are simple to make. From a large piece of tagboard, cut a circle, and then cut a slit in the circle from the edge to the exact middle. Overlap the two resulting edges until you have a long slender cone shape. Then tape the edges together with masking tape.

The hat in Illus. 48 is decorated with pieces of

Illus. 50. See this rare hat in color in Illus. 45.

27

Illus. 51. This Christmas ornament is fashioned from two tissue-paper poinsettia blossoms. Each petal is outlined with one red pipe cleaner, the ends of which are glued in the middle of the flower. Six pipe cleaners radiate up from the petals of each flower and are joined to a piece of ribbon at the middle.

Illus. 52. A beautiful fluffy green tissue-paper Christmas tree will make a festive centerpiece for your holiday table. This tree is constructed on a chicken-wire frame. (See page 35.)

Illus. 53. To make this frilly tissue-paper lamb, see page 31.

Illus. 54. This enormous flower can be turned into a "cheering mop" in no time by using larger and more tissue paper sheets. (See page 34.)

Illus. 55. This group of tissue-paper figures were all fashioned from pieces of rolled tagboard.

Illus. 56. Black tissue paper enshrouds the witch's tagboard body.

tagboard-cylinder body and cylindrical arms and head. She is completely encased in black tissue paper.

Three tagboard cones, covered with tissue make up the body and the wings of the angel on the left in Illus. 57, while the tagboard candle in the middle is covered with water-soaked tissue, as is the holder. On the right is a four-petalled flower cut from tagboard and overlaid with tissue. Pipe cleaners are glued onto the edges of the petals and a paper doily serves as the eye of the flower.

Illus. 57. An angel, a candle, and a flower—all made of tagboard and covered with tissue.

Tissue-Carton Sculpture

Besides simply decorating containers and cartons with flat tissue such as you have done up to now, you can combine various cartons by joining them with tape or staples and apply tissue in such a way that no one would ever guess what was underneath! For example, would you believe

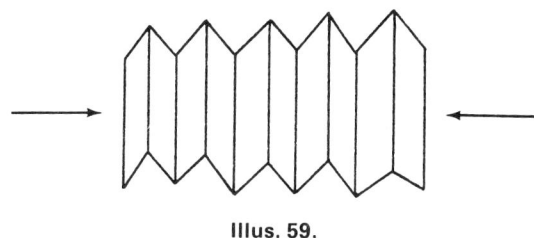

Illus. 59.

that a salt box and a toilet-paper roll lurk under that frilly lamb in Illus. 58?

The salt box forms the body of the lamb and the paper roll the neck. The cardboard rolls often found on coat hangers have been used to make the lamb's legs. In addition, plastic foam has been combined with the carton material. The head is a hard plastic-foam (Styrofoam) ball painted with tempera, and the nose is a smaller ball cut in half and glued onto the head.

The entire body is camouflaged by tissue-paper "flowers." To create a flower, take a piece of colorful tissue, approximately 4" × 8", and fold it back and forth in an accordion pleat as shown in Illus. 59. Gather the accordion pleats together at the middle (arrows) and tie tightly with thread, and you will have a compact floral ball. Make as many as you need to cover the lamb and attach them to the body with undiluted white glue. Do this in sections, allowing the glue to dry before going on to another section, in order that the tissue won't be accidentally knocked off. Construction paper ears complete this little figure.

Illus. 58. See this accordion-pleated-tissue lamb in all its colorful glory in Illus. 53.

Illus. 60. Tissue paper is ideal for "stained-glass" window transparencies (page 40). Here, an orange sun radiates its light upon a magenta and purple sky. Green and yellow blades of grass form the lower part of the design.

Illus. 61. Yarn, pipe cleaners, tissue paper and liquid bleach combine to make this Oriental-looking window mobile.

Illus. 62. One of the most striking transparencies you might make is an insect window hanging. Here an imaginative fly and a fanciful bee were cut from construction paper. The open areas were then lined with tissue in the same way as in making a "stained-glass" window. (See page 42.)

Flowers and "Mops"

The exciting colors of tissue paper just beg to be shown off as artificial flowers, and the delicate quality of the tissue itself duplicates the nature of real flowers.

To make the flower in Illus. 63, take a long strip of tissue, at least 18″ × 6″. Fold the long way once so that it measures 18″ × 3″. With the fold at the bottom and the edges at the top, encircle a wire stem with the strip as shown in Illus. 63a. Then wind florist's wire round the base of the

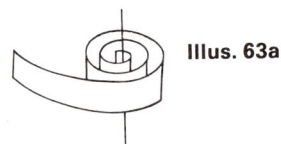

Illus. 63a.

Illus. 64.

Illus. 64a.

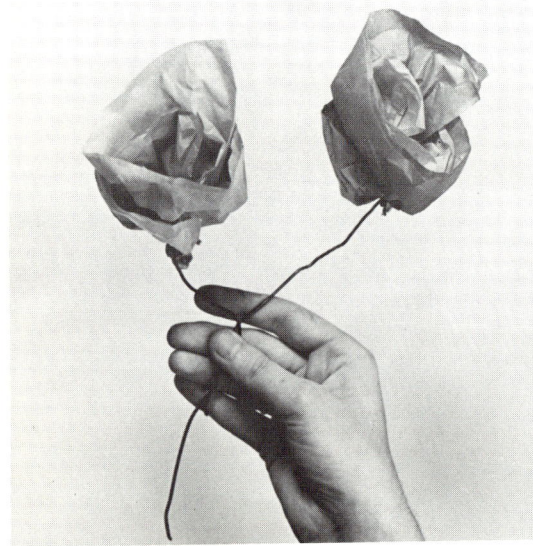

Illus. 63.

flower to draw the bottom towards the wire stem. Fluff up the tissue as shown.

The full, flowing blossom in Illus. 64 is just as simple to make. Cut a piece of tissue approximately 12″ × 24″. Then cut slits into the top and bottom edges as shown in Illus. 64a. Wind the middle part of the strip (the un-slitted part) round a wire stem, and secure with florist's wire. The cut portions should protrude above and below the gather, creating a full flower. You might use more than one sheet of tissue paper in order to create an even fuller flower. Use larger sheets to create enormous "cheering mops"!

(See color Illus. 54 for still another kind of flower.)

Tissue-Chicken Wire Sculpture

Chicken wire forms an ideal base for your tissue creations. It can be cut and bent into almost any shape you desire. The cutting jaws of a pair of ordinary pliers are sufficient for cutting the wire, although if you have a diagonal wire cutter, it will cut more easily. Chicken wire is also called poultry netting and wire netting, and it comes in various sizes of mesh—generally 1 inch or 2 inches. For most of your creations, you will

Illus. 65.

Illus. 66.

probably want the smaller 1 inch holes. Chicken wire is very malleable and you can make many shapes by merely squeezing or pulling.

In Illus. 65, a Christmas tree is being constructed over a chicken-wire frame. Here a small roll of wire was bent into a cone shape to simulate a tree. To tissue-decorate, simply wad pieces of tissue into balls and insert them into the holes. Make sure you completely stuff the holes so as to cover the wire. You can make animals, dolls, puppets, or anything your imagination dictates with this easy method.

Illus. 67. This window transparency is constructed of tissue sheets with a great variety of shapes cut out. When the sheets are overlapped and glued together, the cut-out areas are filled with new colors. The entire design was glued onto a border of black construction paper.

Illus. 68. You can create lovely gift-wrapping paper using the bleach-painting technique described on page 43. Here a floral design was bleach-painted on tissue paper.

Illus. 69. A totally different floral design was here bleached onto a sheet of brilliantly colored orange tissue.

Tissue-Wire Sculpture

Using light-weight wire as a base for your tissue creations, you can achieve great variety of forms. Using a single strand of wire, you do not have to cut it or join pieces—simply *bend* it into shape. In Illus. 70, a single piece of wire was bent into the shape of a fish, leaving the two ends open at the mouth. Then white glue was spread on the wire and a large sheet of tissue laid on it to dry. After drying, the tissue was trimmed down to within 1 inch of the wire to form a fish shape corresponding to the wire. Tissue bands and glitter were glued to the surface.

Combining strands of wire allows you to form more complex sculptures such as the floral arrangement in Illus. 72.

The four Christmas tree ornaments in Illus. 73 are also made of more than one strand. The one on the left is composed of two interlocked circles of wire covered with tissue and adorned with

Illus. 71. This turtle was formed from a single piece of wire in the same way as the fish in Illus. 70. However, in order to form the head, the wire strand was twisted into a circle.

paper doily pieces. The second one is made from two interlocked bell shapes, and the third one from four wire circles covered with tissue. The

Illus. 70. A single piece of light-weight wire supports this large tissue-paper fish.

37

Illus. 72. A number of wire strands, varying in length, were joined to create this "still life." Daisy-like shapes and oblong leaf forms were twisted onto long wire stems. A piece of oil-base clay serves as a planter.

Illus. 73. By combining tissue paper with wire you can create any number of different kinds o Christmas tree ornaments, as you can see here

four circles are actually made of two pieces c wire, each twisted to make two circles, and hel together with a pipe cleaner. The ball on the fa right is made of two interlocking circles.

In Illus. 74, a partridge in a pear tree hangs o the left, adorned with real feathers and tissue paper strands. The angel on the right is formed c a circular wire head and a triangular wire body covered with tissue. The wings are not wir supported—they consist of multi-layers of colore tissue folded into triangular shapes and staple onto the head.

Illus. 74. The traditional "Partridge in a Pe Tree" here joins forces with an angel to adorn Christmas tree.

Transparencies

Mobiles

Hanging near a window, a tissue-paper mobile provides an exciting way to capitalize upon the transparent qualities of tissue. In Illus. 75, arrow-shaped outlines, cut from tagboard, hang from yarn strands suspended from a delicately balanced horizontal wire frame. Sunlight illuminates the

Illus. 76. The interesting design on this tissue mobile was achieved by bleach-painting (page 43).

centers of each arrow which consist of tissue glued to the outer framework of the shapes.

Triangular shapes fashioned from pipe cleaners

Illus. 75. A tissue-paper mobile when hung in front of a window not only provides exciting movement, but color also!

39

"Stained-Glass" Windows

The stunning stained-glass window in Illus. 78 is actually made of tissue! You will be amazed at how sunlight or even strong daylight will transform the tissue into a blazing array of colors.

Illus. 77. A three-dimensional canopy supports the hanging triangles of this mobile.

are attached to strands of black yarn in Illus. 76. Tissue was glued to the pipe cleaners and trimmed after the glue had dried. The unusual design on the triangles was achieved by painting liquid bleach onto the tissue (see page 43).

Illus. 78. A tissue-paper "stained-glass" window

40

To begin, fold a sheet of 12" × 18" black construction paper in half the long way as shown in Illus. 79. After you have decided on your design (make your first one as simple as possible), lightly draw in the "lead tracings" as shown in the illustration. On real stained-glass windows the tracings are all a uniform width, so keep your tracings to approximately $\frac{1}{2}$ inch, which is in proper proportion to a window of 12" × 18". Then cut out the "glass" areas, being sure to cut through both thicknesses of paper, and remove them, leaving only the tracings.

When all the shapes are cut out, open the paper and you should have a completely symmetrical design. Choose your tissue-paper colors and place the *folded* skeleton on the tissue. Trace with a pencil around the inside of each removed area onto the color of tissue that you have chosen for each space. Then cut out the tissue shape slightly larger than the pencil line indicates so that the overlapping edges of the tissue can be glued to the underside of the corresponding tracings. Continue until your entire window is covered with colorful tissue.

If you want to make a larger window, tape as many as four or five sheets of 12" × 18" black construction paper together. Butt the edges and

Illus. 79.

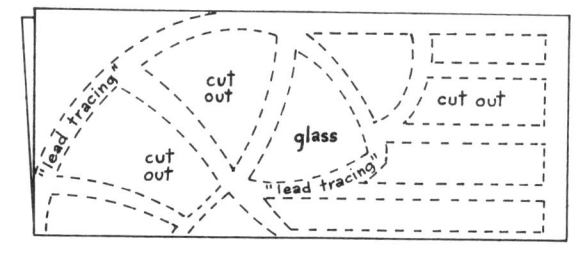

Illus. 80. For removing small shapes, such as here and in Illus. 79, use an X-Acto knife or a single-edged razor blade.

41

Illus. 81. A giant insect window transparency is certain to draw attention!

are hovering in front of the window. These insects are made in the same way as the stained-glass windows. Simply cut out an insect shape from black construction paper and draw a suitable insect design on it. Then cut out the resulting shapes, and glue tissue paper onto the reverse side.

The incredible eight-legged bug in Illus. 81 has touches of Christmas glitter here and there, and in Illus. 82, two delicate creatures, a butterfly and a dragonfly, grace another window.

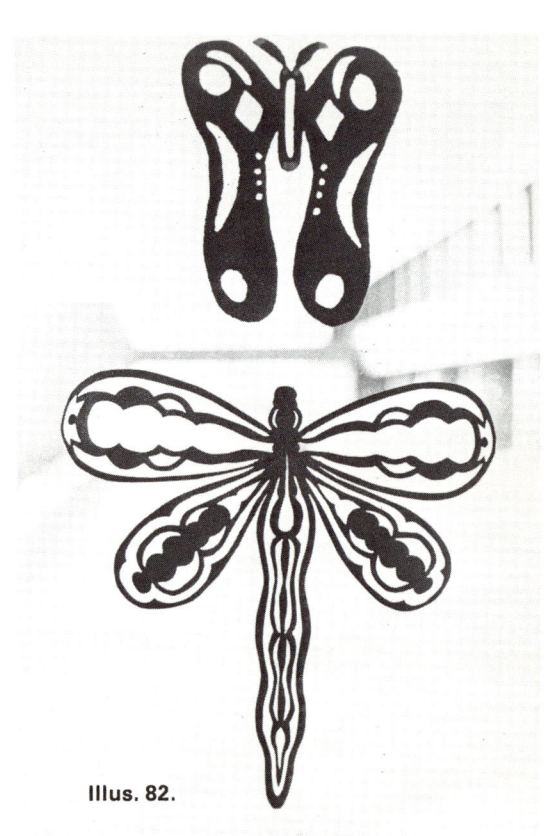

Illus. 82.

join them on the reverse side with masking tape. Fold the joined papers in the middle in the long direction and proceed as before.

Insects

A fantastic transparency is shown in color Illus. 62—two gigantic insects, a fly and a bee—

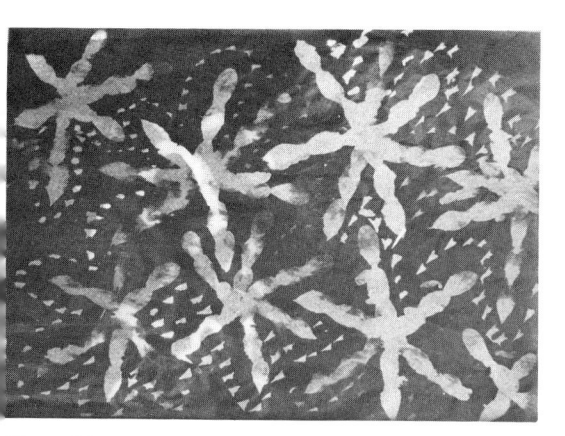

Illus. 83. Liquid bleach created this star-like motif on dark purple tissue paper.

Bleached Tissue Creations

Illus. 84. Various sizes of daisies and flowing lines characterize this attractive bleach-painting.

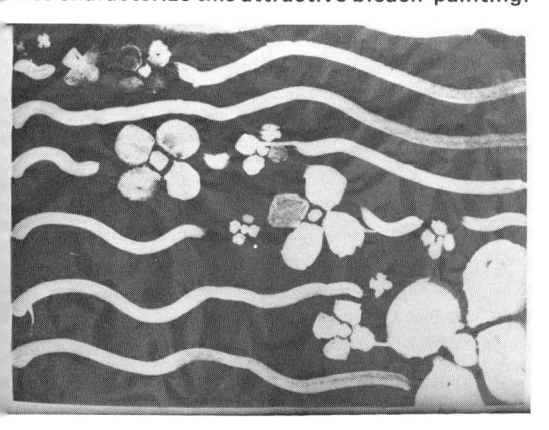

When liquid laundry bleach is applied to tissue paper, it removes the color. You can control the bleaching as you please to create abstract designs or representational ones.

Bleach Paintings

Since liquid bleach could ruin your paint brush, fill a small pliable plastic bottle about three-quarters full of household chlorine bleach. Insert a strip of foam rubber into it so that it touches the bottom and protrudes about $\frac{1}{2}$ inch at the top. For light lines you simply need to run the tip of the rubber across the tissue, and for large areas you can squeeze the bottle.

However, if you want to use a brush, do so. The design in Illus. 83 was created with a bamboo brush. The point of the brush was used to stipple

Illus. 85. Here a representational scene was bleach-painted on tissue. A road, bordered by flowers and trees, leads to a small house.

43

Illus. 86. Bleach painting can be applied to creating unique gift-wrap paper as shown here.

on the small triangular shapes encircling the star motif.

You can use the bleach-painting technique very effectively in making one-of-a-kind gift wrap as shown in Illus. 86. Here, random strokes of the squeeze bottle were applied to dark purple tissue. See color Illus. 68 for another example of a bleach-painted gift paper.

Bleach Printing

In addition to painting bleach on the tissue, you can use the bleach in the same way you would use ink in potato printing.

In Illus. 87, an "X" was cut from a potato. When cutting your stencil, do not simply cut the design in half a potato—cut a whole piece of potato into the shape of your design

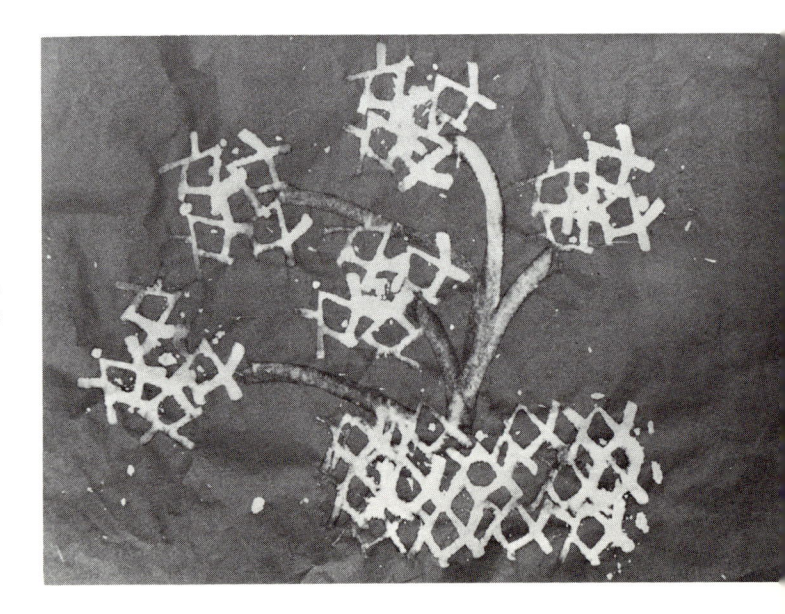

Illus. 87. Instead of using ink in potato printing, you can use bleach!

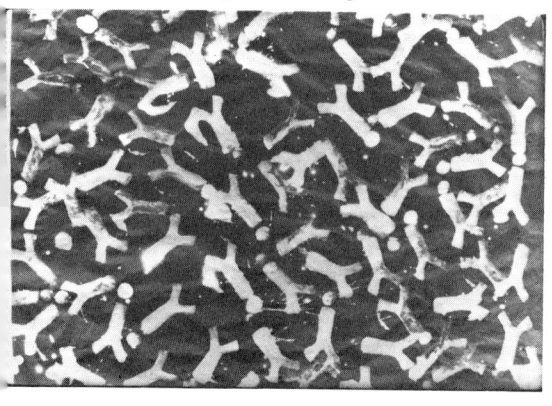

Simply dip the potato into the bleach and press it onto the tissue. The "X" shape in Illus. 87 was used to create a representational design—flowers in a vase.

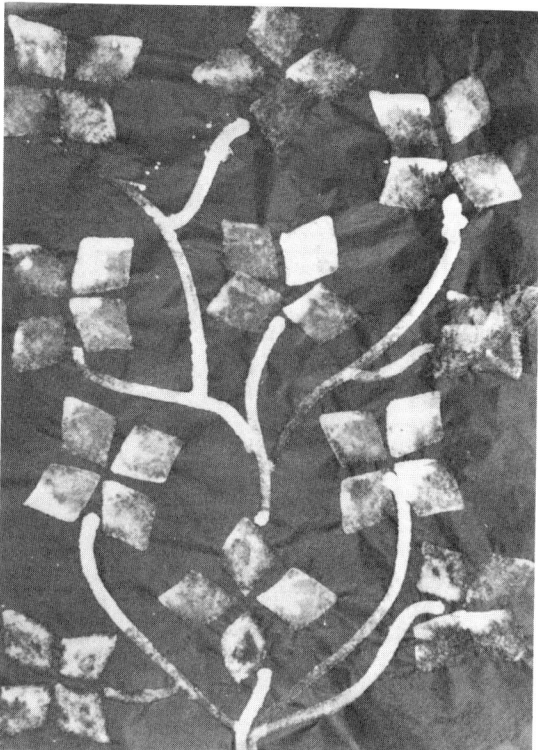

Illus. 88. This octopus-like design was made with stencils cut from a potato and dipped into liquid bleach and then combined with long curved brush strokes.

like this:

and not like this:

Illus. 89. A "Y"-shaped potato stencil created this rhythmic design.

Illus. 90. This floral design was achieved with a diamond-shaped potato stencil and brush strokes.

45

Illus. 91. A tree form cut into a linoleum block produced this forest scene.

Printing on Tissue

Illus. 92. A flat piece of oil-base clay with the veins of a flower pressed into its surface was inked and repeatedly pressed onto the tissue paper.

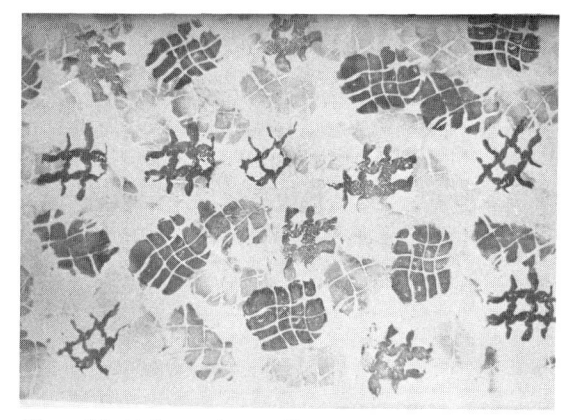

Illus. 93. Using wire, a line design was pressed into a piece of clay to print this motif on tissue.

The vibrant colors of tissue paper make it an exciting medium to print upon. In Illus. 91, various colored shapes of tissue were glued to white drawing paper with a mixture of half

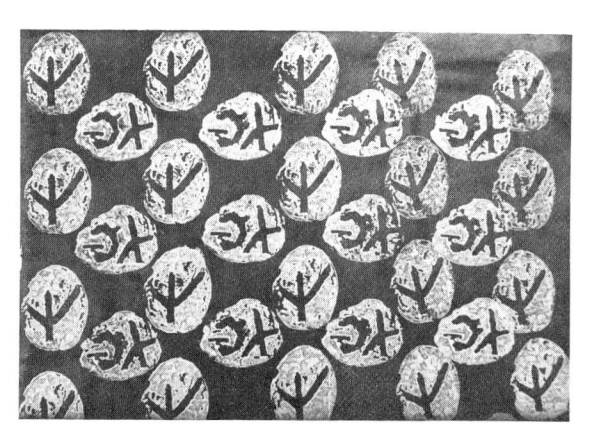

Illus. 94. Here the well-known peace symbol was adapted to potato-print a design.

Illus. 95. This motif was created from a linoleum block with a floral design gouged into its surface.

white glue and half water. The overlapping colors created new hues. A tree form was cut into a linoleum block (with negative background areas cut away) and was used to print the rhythmic design upon the colored "foliage." Battleship linoleum is excellent for linoleum printing, and water-soluble printers' inks are easy to clean up.

In Illus. 92, a flat piece of oil-base clay was used for printing. The flower design was pressed into the clay surface, inked, and repeatedly applied to the tissue paper.

Index

angel, 30, 38
backgrounds, 4
balloons, laminating on, 19, 24
baubles, 7, 20
bleached tissue creations, 43–45
bleach painting, 36, 39, 43, 44
bleach printing, 44, 45
"bleeding," 9, 18, 19, 21
candle, 30
carton-tissue sculpture, 29, 31
"cheering mops," 29, 34
chicken wire-tissue sculpture, 28, 35
Christmas decorations, 7, 28, 30, 35, 37, 38
cityscape, 11
clowns, 24, 55
collage, dye-transfer, 12
collages, laminated tissue, 14, 15, 16, 20
containers, laminated, 15, 18, 19, 21
cord drawings, 4–5
crayon-stencil technique, 22
crushed tissue figures, 6
designs, planned, 12, 13
designs, unplanned, 9, 13

drawings, cord, 4–5
dye transfers, 9–13
Easter eggs, laminated, 19, 24
elephant, 6
figures, 6, 12
finishes, for collages, 15
fish, 5, 13, 37
flowers, 5, 14, 26, 28, 29, 30, 34, 38, 45, 47
gift-wrap paper, 36, 44
giraffe, 24
glass, laminating, 15, 18, 21
greeting cards, tissue-parchment, 8
hats, party, 25, 27
insects, 33, 42
lamb, 29, 31
laminated tissue creations, 14–19, 20, 21, 24
landscapes, 10, 13, 23, 43
letters, laminated, 17, 20
lion, 55
magazines, use of, 17
mobiles, 32, 39, 40
mouse, 6
painting, bleach, 36, 39, 43, 44

paper plates, laminated, 16
photographs, 7, 20
possibilities of tissue, 4
potato printing, 44, 45
printing on tissue, 44, 45, 46–47
sculpture, tissue-carton, 29, 31
sculpture, tissue-chicken wire, 28, 35
sculpture, tissue-tagboard, 24, 25, 27, 30
sculpture, tissue-wire, 37, 38
seascapes, 5, 12, 13, 16
snowflake designs, cutting, 18
"stained-glass" windows, 32, 40–41
stencilled designs, 22–23, 26
tagboard-tissue sculpture, 24, 25, 27, 30
texture in collages, 14, 15
tissue dye transfers, 9–13
tissue-parchment cards, 8
train, 24
transparencies, 32, 33, 36, 39–42
trees, 10, 11, 12, 23, 46
turtles, 14, 37
wire-tissue sculpture, 37, 38
witch, 27, 30